BRIDGING THE TACOMA NARROWS

TEXT AND PHOTOGRAPHS BY JOHN V. ROBINSON

Carquinez Press

First Printing 2007

ISBN 978-0-9744124-6-7 (hardcover)
     978-0-9744124-7-4 (softcover)
LCCN 2007901288

Copyeditor: Deborah Heinzmann

To order additional copies of this book please contact:
Carquinez Press
P.O. Box 571
Crockett, CA 94525
www.carquinezpress.com

Printed in China by Permanent Printing Limited
Design by : Design Action, Oakland CA 94609

The Fund for Labor Culture & History would like to thank the following organizations for their generous support of this project:

• International Association of Bridge, Structural, Ornamental, and Reinforcing Iron Workers.

• Ironworkers Management Progressive Action Cooperative Trust (IMPACT)

• Iron Workers Local 86.

• Pacific Northwest District Council of Iron Workers

• Tacoma Narrows Constructors

CONTENTS

FOREWORD

When I became an ironworker in 1962, I never imagined that one day I would be presiding over the International Ironworkers Union as its General President. I began my career in St. Louis at Local 396, and was always told by my ironworking family: "Work hard and anything is possible." As a third generation ironworker, my connection and dedication to the union runs deep.

Our union derives its strength from its membership. In the more than 110 years since its founding, the Ironworkers International has gone from a staff of volunteer officers without an established headquarters to an international organization with more than 140,000 proud members. Ironworkers have often been called "cowboys in the sky." This colorful metaphor hints at the dangerous job ironworkers are asked to perform daily. Even though our safety standards have improved over the years, the sight of ironworkers working 500 feet up on a new suspension bridge still leaves onlookers breathless.

The completion of the new Tacoma Narrows Bridge marks the third time in 70 years that the Ironworkers have completed a suspension bridge connecting Tacoma to Gig Harbor. The first bridge, nicknamed Galloping Gertie, opened on July 1, 1940 after two years of construction. Gertie was built to sway with the wind, but on November 7, 1940—just four months after opening—Gertie plunged into the Narrows after a strong windstorm. The second bridge, designed to allow wind to pass through it, opened on October 14, 1950 following 29 months of construction. It still stands today, only 185 feet away from its new sibling.

Bridge construction encompasses all aspects of the ironworking craft, beginning with the fabrication of the steel to the tightening of the final connecting bolt. The skilled hands of the ironworkers performing their craft accomplished all of the work. With this type of work there is simply no room for error. Our members take these tasks as a true testament of the skills and traditions that make our trade highly disciplined and unique.

The magnitude of this project only underscores the emphasis the Ironworkers Union places on safety and training. I believe that an injury to one ironworker is one injury too many. This project was not built under sunny skies. The ironworkers toiled day and night under some of the worst weather conditions recorded in Pacific Northwest history.

The new Tacoma Narrows Bridge honors the work of all ironworkers—past, present and future—while bringing pride to our union and to all working people of the Pacific Northwest.

Joseph J. Hunt
Washington, D.C.

6

FOREWORD

For centuries bridges have exercised a fascination on all cultures and civilizations, from tree trunks that fell across brooks, to the drawbridges used to keep enemies at bay; the sense of crossing and reaching the other side; of bringing us closer; of facilitating and encouraging communications; eliminating gaps; and, the almost philosophical concept of opening ourselves to what is to come. Bridges have been immortalized and idealized by painters, poets, lovers and dreamers while, with the passing of time, bridges have stood as silent witnesses to all those that have crossed them in their journeys.

While one may take for granted the way bridges influence our lives or, perhaps, one could not imagine life without them, we are rarely indifferent to their monumental scale, their unique beauty, and in the case of suspension bridges, their sheer simplicity and majestic presence. Bridges inevitably become cultural icons, geographical landmarks and visible points of reference to what human ingenuity can achieve.

Would a bridge be a bridge if no one were there to cross it? Or would it ever be a bridge if no one were there to build it? It is our industrious work and our journeys that allow them to exist. So it is always the human factor that makes the difference. It is the inventiveness, drive, hard work, dedication, the willingness to undertake extraordinary endeavors, the unwavering determination to face and resolve challenges of all kinds and, above all, the individual and personal satisfaction that we have been part of a project that will be the everlasting legacy of our collective effort and work.

This project is more than the parts and pieces that make up its structure. For all the people involved in the building of this bridge, it is a life experience full of satisfactions, emotions, and hardships that are an integral part of the bridge itself.

This book is about the journey of making it happen; about the transformation of dreams and ideas to realities; about the stages of construction but mainly, it is about people: the men and women that built and made the new Tacoma Narrows Bridge

John Robinson tells the story of our journey as only a construction worker turned photographer can do it, leaving a documentary testimonial of significant historical and artistic value.

Manuel Rondón
Tacoma Narrows
Constructors
Project Manager
July 2007

FOREWORD

My earliest memories as a young boy, whose father and three uncles were ironworkers from Local #114 in Tacoma, were of conversations about when a new narrows bridge would be built. As the years went by and I graduated my apprenticeship to become a journeyman ironworker, the new bridge was almost mythical. Though we all knew it was needed there was always debate as to whether it would ever be built.

In 1999 when Iron Workers Local #114 in Tacoma merged with Local #86 in Seattle I was asked to come on staff as a Business Agent. This was when I started to believe that a new narrows bridge could become a reality. One of my first assignments was to represent the Iron Workers in negotiating the Heavy and Highway Project Labor Agreement for the new bridge through the Pierce County Building & Construction Trades Council.

Since that time I have served in several positions working toward the safe and successful completion of this project. As a representative for the ironworkers, I had the opportunity to work with Sen. Bob Oke in lobbying the state legislature for the financing of the bridge and to serve on a task force to mitigate the impacts of foreign steel. As President of the Pierce County Building & Construction Trades Council I helped implement the Project Labor Agreement, and finally as the Executive Secretary for the Washington State Building & Construction Trades Council.

While this is personally gratifying, my greatest sense of pride comes from representing the men and women of the building trades who built this bridge. From the sinking of the caissons to the painting of the bridge trained craftsmen and women did an exceptional job working in extremely harsh and unforgiving weather conditions. They deserve every ounce of respect we can muster. They certainly have mine!

On behalf of the Washington State Building & Construction Trades Council and its affiliates I would like to congratulate Tacoma Narrows Constructors for their outstanding safety record. They have proven that when labor and management work together loss of life and serious jobsite injuries can be eliminated.

I would also like to acknowledge the oversight and guidance of the Washington State Department of Transportation. While working through the complexities of this project Secretary of Transportation Doug MacDonald and his staff always focused on the best possible working conditions for the men and women of the building trades.

We live in an era when few earn a living working out in the wind, rain and cold. This book should remind every reader of the critical need for the men and women of the building trades willing to challenge the elements while building the highways, bridges, skyscrapers and dams, that shape our future.

David D. Johnson,
Executive Secretary
Washington State Building &
Construction Trades Council
AFL-CIO

FOREWORD

Every great bridge springs from the vision and effort of hundreds of people. On the magnificent new Tacoma Narrows Bridge we at the Washington State Department of Transportation got to see them all. As the owner, all the participants came into our field of view.

The ironworkers provided thousands of hours of the work from skilled, dedicated, and hardy men and women. Their work is woven into the bridge from the bottom of the caissons to the top of the towers, from east to west, from one buttressed anchorage to the other.

The rebar tied by ironworkers on this bridge is counted in thousands of tons. The cable they spun is measured in thousands of miles. It would be hard to find a square yard of the bridge that an ironworker didn't touch. But ironworkers were just one of many crafts and trades that built this bridge.

Operating engineers, carpenters, cement masons, laborers, teamsters, electrical workers, plumber/pipe fitters, sheetmetal workers, roofer/water proofers, bricklayers, asbestos workers, and painters all had a hand in this bridge. Almost three million hours were worked at the Narrows as men and women put the bridge together one piece at a time, hour after hour, day after day, season after season.

Then there were the engineers. They designed and planned how to build the bridge. They brought a myriad of disciplines to the task. They spoke in accents and languages from around the world. They created their calculations and drawings; made their surveys and inspections to ensure quality in every aspect of the bridge.

Then there were the business managers. They conceived how the project would be organized and they saw it through right down to the tracking of the construction payrolls, and set-up of the newest technology to automatically and accurately charge the tolls that will pay for the bridge.

Next we come to the public officials. They passed and signed the legislation that allowed the bridge to be built. Others wrote the permits to protect the environment as the bridge was built. Others borrowed on Wall Street on the bridge's behalf the hundreds of millions of dollars that financed the construction as the caissons went into the mud, the towers came out of the water, the cable was spun and the deck sections were lifted. Local officials kept a watchful eye to keep traffic moving and help minimize the construction's effect on the neighborhoods.

The press is ever present for a major infrastructure project, and so it was on this bridge. The public learned of the project through caisson-to-cable media coverage of the techniques of constructing the structure and the challenges overcome by the engineers and trades people.

When all is said and done, it is the people's bridge. At the beginning, some loved it and some hated it and some just worried about the disruption to their commutes and neighborhoods. The progress of construction changed some minds. Everyone was caught up in the drama of the work.

Everyone was captured by the feats of human ingenuity, skill and strength as the bridge was seen rising in our midst. Drivers and passengers rubber-necked as they crossed the old bridge. They watched from the shore and from their boats. They watched from afar over the web cameras and the official Internet site and even on the unofficial websites that sprung up around the project. In truth, all the citizens became fans of the most exciting show in town, the construction drama of the new bridge.

Now we turn the new Tacoma Narrows Bridge over to the citizens, our customers. It is their bridge. Thanks and congratulations to everyone who contributed. It has been a great project and it is a great bridge. Bridge builders will tell you, "It's the experience of a lifetime to work on this." The memories are rich. Like good fish stories, they will get even bigger and better as time passes. The best, however, is yet to come. Years from now our children will drive over the Narrows and say to our grandchildren, "My Dad –or Mom– built this bridge."

Douglas B. MacDonald,
Secretary of Transportation—
July 2007

ACKNOWLEDGMENTS

I wish I had more space to acknowledge all the people who helped make this book a reality. Over the past two years I have met, talked to, and photographed many people. Ron Piksa at the Pacific Northwest District Council of Iron Workers, Steve Pendergrass and the officers of Local 86, all deserve recognition for supporting my project. I especially want to acknowledge Doug Smith for his time and attention to this project over the past two years.

From California I want to thank Archie Green at the Fund for Labor Culture & History for his continuing friendship and support. Dick Zampa, retired First Vice President of the International Iron Workers, deserves my thanks for his continuing support of my many projects and for introducing me to Ron Piksa.

From TNC I want to thank Manuel Rondón, Pat Soderberg, Reg Carson, and Erin Hunter for being gracious hosts on my many trips to the bridge and their support of this project.

Tacoma resident Robert Pottenger has followed the project from the beginning and generously shared some of his photographs with me. Most of all I would like to thank the people who built the bridge. People like: Bob Ashmore, "Cowboy" Butler, Ron Carrier, Brady Cooper, Jim Kostelecky, "Pyro" Thom Schell, "Psycho" Steve Seidel, Richard Sokolik, and many others who encouraged me in my efforts and took the time to explain the work process to me. Were it not for the help of such people this book might not have been possible.

John V. Robinson
May 2007

INTRODUCTION

Bridges unite people and communities and are among our most significant architectural achievements. People involved in building a great bridge never forget their experiences. With two bridges already a part of architectural history and legend, "Galloping Gertie" and its still-standing replacement "Sturdy Gertie," the bridges across the Tacoma Narrows are on a par with the Brooklyn Bridge and the Golden Gate Bridge.

With such a rich history it is only natural that retired ironworkers would take an interest in the new bridge and talk about their experiences on the first two bridges. They meet for lunch once a month at the Pine Cone Café in Tacoma. They come to talk about their lives, and talk about the first two bridges, and talk about the progress of the new bridge.

Many books have been published celebrating bridges as architectural monuments. Few books document the con-struction of our important bridges, and fewer still have doc-umented the contributions of the engineers, piledrivers, ironworkers, carpenters, labors, electricians, teamsters, and crane operators, who design and build our great bridges.

Much has been written about the 1940 failure of "Galloping Gertie." In 1990 Joe Gotchy, a retired operating engineer, published *Bridging the Narrows* wherein he recounts his experience working on the first two bridges across the Tacoma Narrows. Gotchy's book is a rare contribution by a worker to our understanding of the first two bridges. In 2006 Washington State University Press published Richard Hobbs' *Catastrophe to Triumph: Bridges of the Tacoma Narrows* a first rate scholarly history of the first two bridges with hundreds of historic photographs that detail the events that lead to the "Galloping Gertie" disaster and the rise of its successor.

Retired ironworkers meet at the Pine Cone Café. From left to right are: Rudy Lutz, Gordon Bolen, Ben Freudenstein, Ken Haygett, Bill Matheny, Will Hyduke, Gordy Irwin, Earl White, and Earl "skeeter" Bachman. When I told Earl White I was writing a book about the new bridge he looked at me wryly and said, "Make sure you put in that book that Tacoma Local 114 built those first two bridges." Their connection to the first two bridges and their old local, which in 1999 was absorbed into the Seattle Local 86, runs deep

All new bridges generate a certain amount of political, social, and economic controversy: "It's too big!"…"Not in my back yard!"…"It's too expensive!"…"and the ever-present cry of the sidewalk superintendent… "It's impossible!" The current book, *Bridging the Tacoma Narrows* (the title pays homage to Gotchy's book), is a preliminary document of the construction of the new Tacoma Narrows Bridge—I'll leave it to posterity to explore the social, political, and economic implications of the new bridge.

This book has been a cooperative effort between the Ironworkers and Tacoma Narrows Constructors. I took the majority of the photographs; a few photos were contributed by interested bystanders who have followed the bridge's construction since the start. The contractor, TNC, provided details about the bridge's construction and other textual elements that I have edited into the text.

Ultimately, this book is a series of judgement calls. I have tried to make the descriptions plain, readable—without being overly technical—and tie them together with the photographs. This book is primarily a photo essay. I have also tried to achieve a balance between the technical and aesthetic aspects of the bridge and to allow readers to experience and appreciate the various aspects of the project.

The bridge itself seems immense and complex. Yet it consists of only five major elements: the caissons that support the entire bridge, the towers that support the main cables, the main cables that support the road deck, and the road deck itself. With this in mind, I have divided the book into chapters that more or less follow this simple yet elegant template. My only additions are a chapter on constructing the footbridge that allowed workers access from shore to shore and tower to tower while the bridge was being constructed and, the last chapter, which is a portfolio of photographs that capture some of the abstract beauty of the bridge as well as some of the human toil involved in making this magnificent bridge. I hope you like the result.

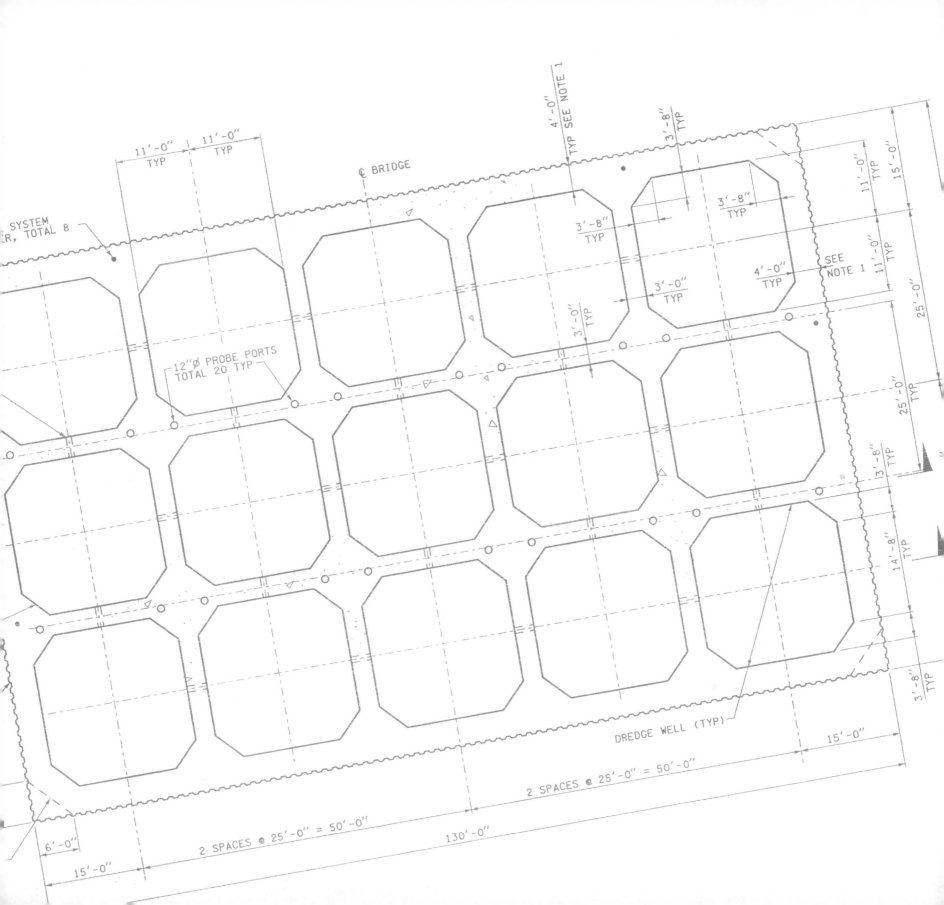

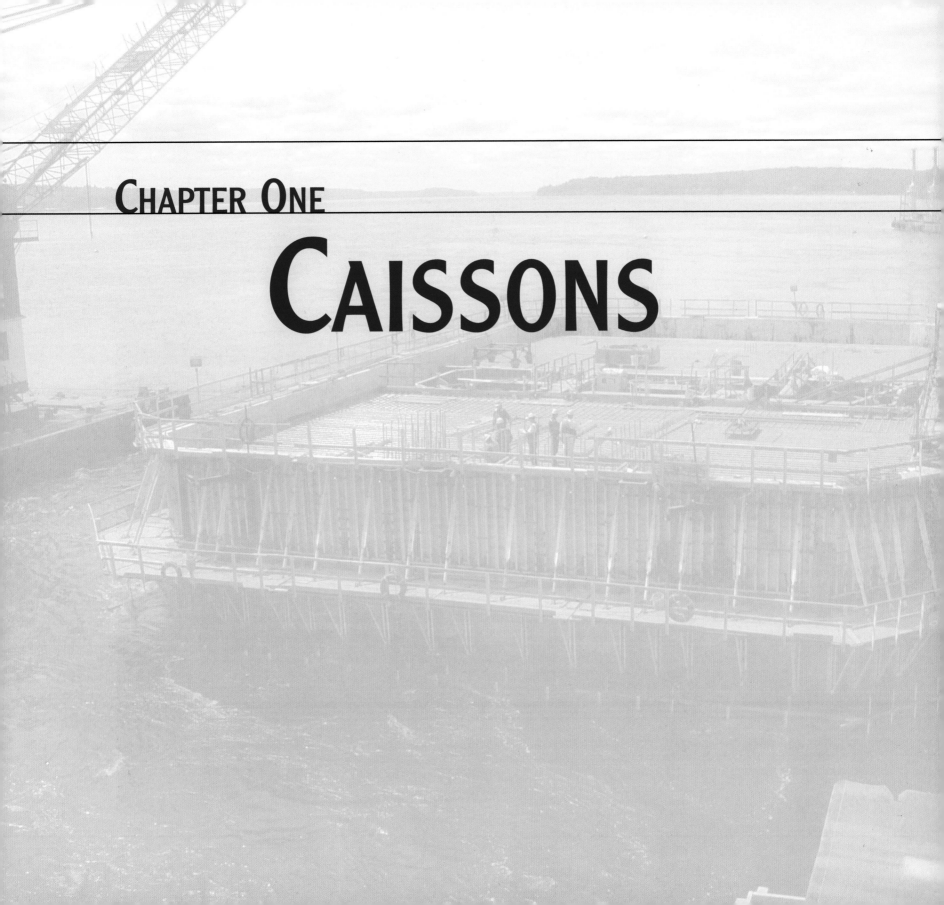

CHAPTER ONE
CAISSONS

One of the most important structural elements of the new bridge is unseen, namely the foundations for the two bridge towers. Called "caissons" they were also one of the most technically challenging aspects of the bridge's construction. There are several preferred methods for constructing bridge foundations. If the geology is favorable piles can be driven to bedrock and the bridge footing can be built on top of the piles. If the water is shallow and placid cofferdams can be used. A cofferdam is an enclosed area with the water pumped out — essentially it's a hole in the water that workers can descend into to construct the bridge foundation and footing. The third method, and the one used at the Tacoma Narrows Bridge, is the open dredge caisson. In this method a small section of a floating box caisson is positioned in the water above the sea bottom. The bottom of the caisson has a steel cutting edge that helps it penetrate through the mud and overburden until it reaches bedrock. As more of the caisson is built-up on top, the weight causes it to slowly sink to the bottom of the sound.

The cutting edge of the caisson was built at Todd Pacific Shipyards in Seattle in early 2003. When the cutting edges for the two caissons were completed they were towed to the Port of Tacoma, closer to the bridge site, for further construction of the outer steel skin. By the summer of 2003 the two steel and concrete boxes, which measured 130 ft long, 80 ft wide and 78 ft high and weighed in at 14,000 tons,

Looking down from on the Gig Harbor caisson from the old bridge. Green epoxy coated rebar is on the supply barge and derrick barge is between the caisson and the supply barge.

were towed to their respective positions at the bridge site. When completed, each caisson weighed in at 85,000 tons.

Once the caissons arrived on site, workers spent the summer and fall of 2003 building up the inner and outer walls of the caisson as it slowly sank to the bottom. The interior of the caisson consisted of fifteen octagonal dredge wells about 22 ft by 22 ft each with an air chamber at the bottom. The combination of air pressure and water allowed into the dredge wells gave a great deal of control to engineers as the caisson sank to the bottom.

It took roughly six months of adding rebar and concrete to the caissons before they touched down on the bottom—Gig Harbor in December 2003 and Tacoma in January 2004.

Once the caissons touched down, divers removed the air chambers and the workers continued building the caissons walls. As the mud was dredged out through dredge wells, the caissons sank roughly 60 ft below the Narrows mud line before settling on bedrock, a firm perch for the two massive towers to come.

Once the caissons were firmly on bedrock the dredge wells were filled with concrete and by mid-summer 2004, bridge workers covered the top with rebar and poured a 15 ft concrete distribution cap. The distribution cap became the footing for the new bridge towers.

GIG HARBOR CAISSON
West Narrows Water Depth: 133 feet
Depth below Seabed : 57 feet
Total Height of Finished Caisson : 190 feet

TACOMA CAISSON
East Narrows Water Depth: 154 feet
Depth Below Seabed: 62 feet
Total Height of Finished Caisson: 216 feet

Ironworkers construct caisson's interior rebar wall.

17

CHAPTER TWO

ANCHORAGES

Two major forces are at work on a suspension bridge: tension and compression. The towers of the bridge are in compression as they support the great weight of the main cables and the road deck as well as the live loads applied to it by traffic and wind. The main cables are in tension, held in place at both the east and west shore by the anchorages.

Each concrete anchorage is roughly 150 ft long, 116 ft wide, weighs in at roughly 81 million pounds and is rooted more than 60 feet into the earth. The anchorages, as their name implies, act as counterweights holding the main cables firm by a combination of their great weight and their friction against the ground.

Even though each of the two anchorages contain 1.4 million pounds of reinforcing steel and 20,000 cubic yards of concrete they are not solid. They are hollow. Atop each anchorage are two large splay saddles that cradle the 20.5 inch main cable as they come down into the anchorage and divert the cables down into a large cement cavern called the "splay chamber." Inside each splay chamber the 19 compacted strands that make up the main cable are separated and looped around 19 horseshoe-shaped strand shoes. Each strand shoe is attached with two large bolts embedded deep in the back face of the splay chamber. In this manner the enormous pull of the main cables are distributed across a broad area at the back of the anchorage.

By May of 2005 the massive Gig Harbor Anchorage was complete and the splay saddles were in place.

To provide the massive amount of concrete required for the job, a concrete batch plant was constructed on site.

Ironworkers on the A-frame that would support the tram-strands for the cable-spinning. The splay saddle is beneath the A-frame.

25

Looking through the Tacoma splay saddle at the bird cages atop the east tower.

Strand shoes staged near the Tacoma anchorage.

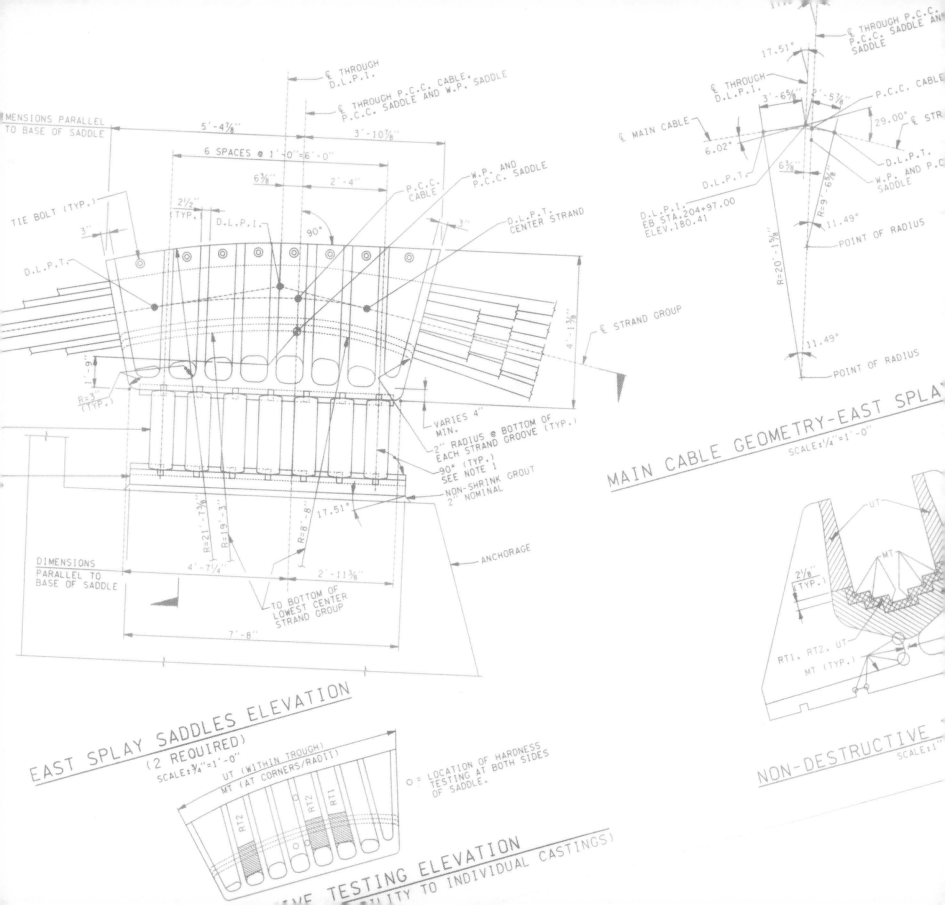

DIMENSIONS PARALLEL
TO BASE OF SADDLE

€ THROUGH
D.L.P.I.

€ THROUGH P.C.C. CABLE,
P.C.C. SADDLE AND W.P. SADDLE

5'-4⅞"

3'-10⅞"

6 SPACES @ 1'-0"=6'-0"

6⅜"   2'-4"

W.P. AND
P.C.C. SADDLE

2½"
(TYP.)

P.C.C.
CABLE

TIE BOLT (TYP.)

3"

D.L.P.I.

90°

3"

D.L.P.T.
CENTER STRAND

D.L.P.T.

R=3"
(TYP.)

1'-9"

€ STRAND GROUP

4'-1⅜"

VARIES 4"
MIN.

2" RADIUS @ BOTTOM OF
EACH STRAND GROOVE (TYP.)

90° (TYP.)
SEE NOTE 1

NON-SHRINK GROUT
2" NOMINAL

DIMENSIONS
PARALLEL TO
BASE OF SADDLE

R=21'-7⅜"

R=19'-3"

R=8'-8"

17.51°

ANCHORAGE

4'-7¼"

2'-11⅜"

TO BOTTOM OF
LOWEST CENTER
STRAND GROUP

7'-8"

EAST SPLAY SADDLES ELEVATION

(2 REQUIRED)
SCALE:¾"=1'-0"

UT (WITHIN TROUGH)
MT (AT CORNERS/RADII)

○ = LOCATION OF HARDNESS
TESTING AT BOTH SIDES
OF SADDLE.

RT2   RT2   RT1

...VE TESTING ELEVATION
...LITY TO INDIVIDUAL CASTINGS)

€ THROUGH P.C.C.
P.C.C. SADDLE AN...
SADDLE

17.51°

€ THROUGH
D.L.P.I.

3'-6⅝"   2'-5⅞"

P.C.C. CABLE

€ MAIN CABLE

6.02°

29.00° € STR...

D.L.P.T.

6⅜"

D.L.P.T.

W.P. AND P.C...
SADDLE

D.L.P.I.
EB STA.204+97.00
ELEV.180.41

R=9'-6⅝"

11.49°

POINT OF RADIUS

R=20'-1⅝"

11.49°

POINT OF RADIUS

MAIN CABLE GEOMETRY-EAST SPLA...

SCALE:¼"=1'-0"

UT

MT

2⅛"
(TYP.)

RT1, RT2, UT

MT (TYP.)

NON-DESTRUCTIVE

SCALE:1"...

CHAPTER THREE
TOWERS

For many people the new bridge became a visible reality when, in the summer of 2004, two tower cranes began to rise above the water. This signaled the beginning of the construction of the two towers, shepherded by two yellow Liebherr 550 HC tower cranes. The cranes were capable of jacking themselves up as the two bridge towers rose steadily higher. When the towers topped out to their full 510 ft elevation the twin cranes stood 600 ft above the water. The cranes' jibs reached out 162 ft from their towers and they were capable of picking loads as heavy as 22 tons.

TOWER CRANE

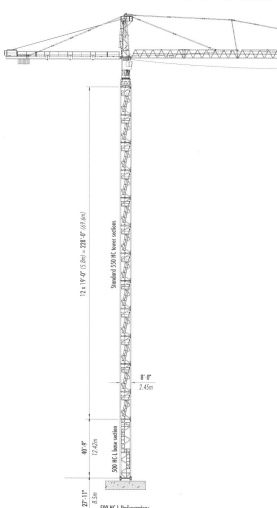

The bridge towers rise 510 ft above the water. Their job is to support the main cables and distribute the weight of the entire structure to their foundations and to the bedrock 200 ft below the water's surface. Each tower consists of two vertical legs that taper in slightly and three cross struts—one below the road deck and two above with x-bracing cast into the concrete to visually simulate the look of the old bridge. The north leg has a series of stairs and ladders that run from the pier top to the top strut. The south leg has an elevator the runs from the pier to the top strut. The top strut has a passage between the tower legs and a manhole to allow access to the tower top for maintenance and inspection of the main cables and the saddle caps.

The towers were constructed using a jump-form method. Each tower leg was surrounded by a movable box form system called the "bird cage." The bird cage weighed roughly 100,000 pounds and was 39 ft x 26 ft wide by 46 ft high. The bird cage was supported by a series of core bolts embedded in the concrete as each new section of tower was built. After a 17.5 ft section of rebar was installed, the concrete was poured and cured, a series of electric jacks pulled the structure up and the process was started again. While one leg was curing the crews moved to the other leg to install the rebar and pour concrete on that side. In this alternating manner the tower legs were constructed in twenty-eight 17.5 ft concrete pours over the course of 10 months.

West tower at about 480 ft in April of 2005.

Bird cage form system from the level of the road deck of the existing bridge. Photo Credit: Robert Pottinger.

A steady supply of concrete was so important to the job that it was mixed on site at a large batch plant set up near the west anchorage. Concrete trucks loaded up and made their way across the old bridge for pours on the east side of the Narrows. On the west side the trucks only had to drive about 200 yards to reach the site. Each truck held about 10 yards of concrete which was delivered to electric pumps sta-tioned beneath the existing bridge. The pumps then forced the concrete through 2,000 ft of 6-inch pipes called "slick-lines" that ran along an access walkway beneath the existing bridge. Once the concrete got to the pier it was pumped into a bucket and hoisted into place by the tower crane. In this manner the towers were constructed one 4-yard bucket at a time.

A cross bridge allowed easy access between the two bird cages. Here Ironworkers Chad "Hippie" Lyons, Troy Birch, Dan Weed, and Kelly Yarnell pose on the cross bridge before heading down for lunch.

On May 19, 2005 the rebar cap (called the "wedding cake" by bridge workers) was lifted into place on the west tower. The piece is adorned with an American flag in a manner of a "topping out" ceremony—a time honored tradition in the construction industry.

Night shift workers set a steel frame over the saddle cap on the west tower. This frame was used to support the tram strands and cable spinning gear as the spinning wheel passed over the saddle cap.

Night shift from left to right: Archie Graham, Gerald Kerr, Daniel Wright, Dennis Shurlock, and Kyle Cooper at the end of their shift.

CHAPTER FOUR

BUILDING
THE FOOTBRIDGE

Before the aerial cable spinning could begin workers had to construct the footbridges, called catwalks, from which to work while spinning the main cables, installing cable bands and suspender cables, and accessing the gantry cranes that would lift the deck sections into place.

Each catwalk consisted of six 1-1/4-inch steel cables, called "floor strands," set about two feet apart. Roughly three feet above the floor strands were two 3/4-inch hand strands. The catwalks followed the same profile as the main cables but were set about three feet lower to give workers easy access during cable spinning and other operations.

In early August of 2005 construction began on the two one-mile long temporary catwalks. The first catwalks were built from the west tower to the Gig Harbor shore. To build the catwalks workers needed a way to get materials between the towers and from the towers to the shore. A series of winch lines was set up on the shore and on the tower tops.

Between August and early October, crews built two one-mile long suspended walkways from the anchorages on shore to the tower tops, draping elegantly across the waterway. Bridge crews used the walkways as work platforms. Made of steel wire mesh and wooden slats separated about two feet apart, the catwalk resembled a wooden footbridge across a jungle ravine one might see in an old movie.

The basic unit of the catwalk consisted of prefabricated 100 ft rolls of wire mesh about the consistency of Cyclone fence. (Indeed, on earlier suspension bridges, such as the Verrazano-Narrows Bridge in New York, Cyclone fence material was used to construct the footbridge.) Wooden slats were wired to the mesh at about two-foot intervals to provide stability and aid traction as the incline was very steep near the tower tops.

By April of 2005 rolls of footbridge material are neatly staged and awaiting deployment to the bridge. In the distance the towers and bird cages are nearing their final height.

To support the catwalks large steel frames were installed at the anchorages and atop each tower. About 15 ft above the floor strands two tram stands were installed to support the tram frames that were positioned at regular intervals along the footbridge to support cable spinning operations. Along with the spinning gear a series of cable formers were attached to the catwalk at regular intervals to keep the wires in their proper place during cable spinning. After cable spinning ended the tram strands supported work carts used to trundle cable bands and suspender cables back and forth across the span.

This shot from October 2005 shows the complete catwalks with the tram frames on place to guide the spinning wheels in their journey across the Narrows. Also visible are one of the nine cross bridges that allowed workers easy access between the two catwalks. The eight large cables beneath the catwalks are holdback cables that ran from the anchorages to the tower tops. As the holdback cables were tensioned the towers were pulled slightly out of plumb towards each shore. As the towers gradually absorbed the weight of the main cables they settled back to vertical. Once the towers were holding enough weight to sit plumb the holdback cables were detensioned and removed.

Ironworkers on the north catwalk of the west side-span. One of the two side-span cross bridges is in place.

Flying up one of the cross-bridges to waiting ironworkers on the catwalk above.

Ironworkers attach a cross-bridge to the floor strand cables as a tugboat and barge pass below.

Ironworkers load a roll of
floor mesh on the Tacoma
side-span catwalk.

Ironworkers on the main span catwalk secure the floor mesh with wire ties.

Art Astala atop the tower frame loading spools of mesh for the catwalk's side panels. The completed catwalk created a safe workspace for workers and kept tools and other materials from falling into the water.

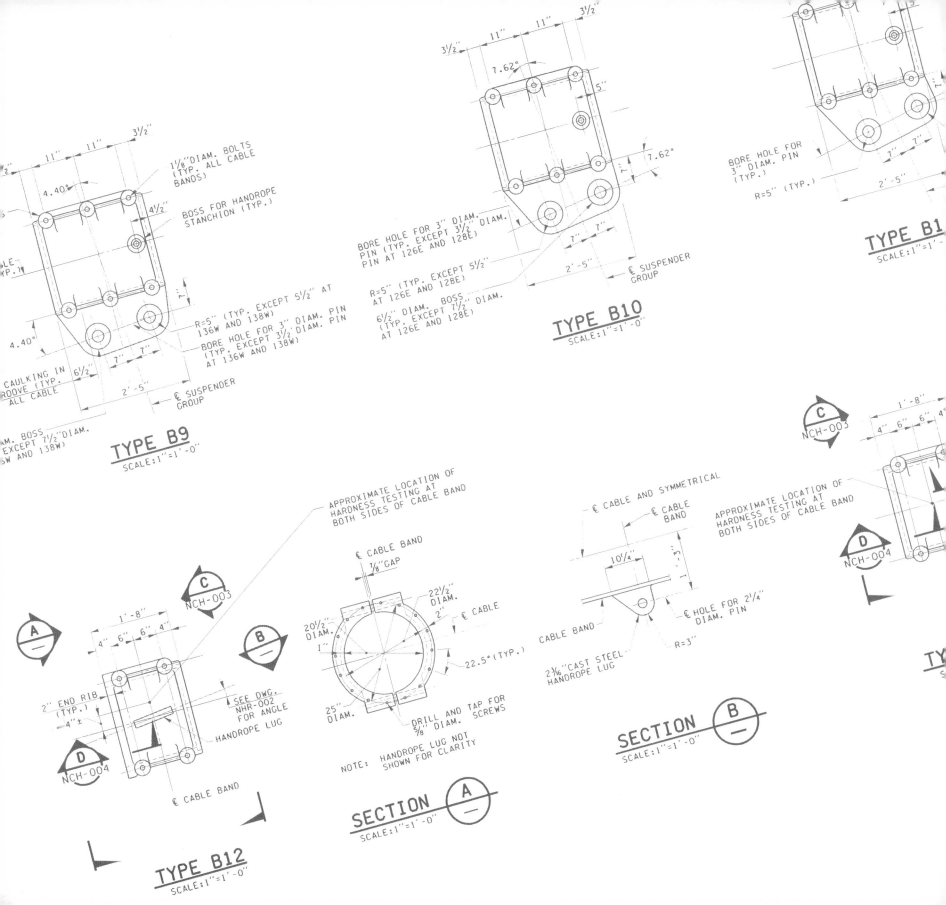

TYPE B9
SCALE:1"=1'-0"

1⅛"DIAM. BOLTS
(TYP. ALL CABLE
BANDS)

BOSS FOR HANDROPE
STANCHION (TYP.)

3½"

11" 11"

4.40°

4½"

7"

R=5" (TYP. EXCEPT 5½" AT
136W AND 138W)

BORE HOLE FOR 3" DIAM. PIN
(TYP. EXCEPT 3½" DIAM. PIN
AT 136W AND 138W)

CAULKING IN
GROOVE (TYP.
ALL CABLE

4.40°

7" 7"

6½"

2'-5"

Ⅽ SUSPENDER
GROUP

AM. BOSS
EXCEPT 7½"DIAM.
W AND 138W)

TYPE B10
SCALE:1"=1'-0"

3½"

11" 11"

3½"

7.62°

5"

7.62°

7"

7" 7"

BORE HOLE FOR 3" DIAM.
PIN (TYP. EXCEPT 3½" DIAM.
PIN AT 126E AND 128E)

R=5" (TYP. EXCEPT 5½"
AT 126E AND 128E)

6½" DIAM. BOSS
(TYP. EXCEPT 7½" DIAM.
AT 126E AND 128E)

2'-5"

Ⅽ SUSPENDER
GROUP

TYPE B1
SCALE:1"=1'

BORE HOLE FOR
3" DIAM. PIN
(TYP.)

R=5" (TYP.)

7" 7"

2'-5"

TYPE B12
SCALE:1"=1'-0"

C
NCH-003

1'-8"

4" 6" 6" 4"

A

B

2" END RIB
(TYP.)

4"±

D
NCH-004

SEE DWG.
NHR-002
FOR ANGLE

HANDROPE LUG

Ⅽ CABLE BAND

APPROXIMATE LOCATION OF
HARDNESS TESTING AT
BOTH SIDES OF CABLE BAND

Ⅽ CABLE BAND

⅛"GAP

22½"
DIAM.

2"

Ⅽ CABLE

20½"
DIAM.

1"

25"
DIAM.

22.5°(TYP.)

DRILL AND TAP FOR
⅝" DIAM. SCREWS

NOTE: HANDROPE LUG NOT
SHOWN FOR CLARITY

SECTION A
SCALE:1"=1'-0"

Ⅽ CABLE AND SYMMETRICAL

Ⅽ CABLE
BAND

APPROXIMATE LOCATION OF
HARDNESS TESTING AT
BOTH SIDES OF CABLE BAND

10¼"

1'-3"

CABLE BAND

Ⅽ HOLE FOR 2¼"
DIAM. PIN

2³⁄₁₆"CAST STEEL
HANDROPE LUG

R=3"

SECTION B
SCALE:1"=1'-0"

C
NCH-003

1'-8"

4" 6" 6"

D
NCH-004

TY
S

Chapter Five

Main Cables

Cable spinning got underway in mid-October when crews pulled the first galvanized steel wire from the Tacoma anchorage over the towers to the Gig Harbor anchorage and back again. The continuous (and spliced) steel wire made 2,204 roundtrips as crews spun the 19,000 miles of cable wire.

Cable spinning is something of a misnomer since the only thing that spins is the wheel that pulls the wire back and forth across the bridge. The process of spinning wire is a simple mechanical process, essentially unchanged since the 19th century.

The wire for the cable, roughly the diameter of a No. 2 pencil, was delivered to the Tacoma anchorage in rolls containing about four miles of wire each. Six rolls of wire were then loaded onto great spools containing roughly 24 miles of wire. The end of each roll of wire was spliced to the beginning of the next roll of wire with a two-inch metal collar called a "ferrule." The ferrule was slipped over the ends of the two wires to be spliced and crimped down with a hydraulic press.

The spinning wheels, two on the north cable and two on the south cable, were attached to an endless tram-cable that ran from anchorage to anchorage. One wheel started in the east anchorage and the other wheel started in the west anchorage. The cable wires were fed through towers on the Tacoma side designed to control the amount of tension on the wires as the wheel pulled them across the narrows.

Looking west at spools of wire and the tensioning towers at the Tacoma anchorage, March 2006

The spinning wheels each had four grooves, and four loops of wire (called "bights") were sent across the bridge at a time. To start the process the ends of four wires were secured at the Tacoma anchorage and the loops were placed in the spinning wheel creating eight parts of wire, four rolling off the bottom of the wheel and four paying out to the top of the wheel from the drums in the Tacoma anchorage. As the wheel made its way across the bridge it pulled eight parts of wire with it.

Ironworker in the east anchorage prepares the spinning wheel for another trip across the narrows, October 2005.

As the wheel moved along the cable's path the four wires on the bottom, called "dead wires" because they were not moving, fell into the saddle caps and cable formers stationed across the bridge. The top four wires, called "live wires" because they were moving, fell in "live wire sheaves" that allowed the moving wires to freely pay out on their journey across the bridge.

Four parts of wire in the live wire sheaves which are paying out from a drum in the Tacoma anchorage. The four dead wires fell directly in the saddle cap as the wheel passed over the tower top. The wires were color coded: yellow, green, black, and red, to make sure each was in its proper place in the strand, October 2005.

Once the wheel reached the west anchorage the bights of wire were removed from the wheel and looped over a steel strand shoe attached to two bolts rooted deep into the back face of the splay chamber. There are 19 strand shoes in each anchorage. As the name suggests, the splay chamber allowed the 19 strands of the cables to splay out and distribute the pull of the main cable across the entire back wall of the anchorage.

While the bights of wire were being secured to the strand shoe in the west anchorage the second wheel, now in the east anchorage, was loaded with four more bights of wire ready to bring another eight parts of wire across to Gig Harbor. When both wheels were ready the tramline was reversed and the two wheels made their way back across the bridge, crossing at mid span. A good way to visualize the process is to think of a ski lift, with spinning wheels attached the tramlines instead of ski chairs.

When the wheel in the west anchorage made its return trip the only thing it did was pick up the wires previously placed in the livewire sheaves and drop them, now dead, i.e. unmoving, into the cable formers and saddle caps. On each trip back and forth across the bridge eight wires were added to the cable strand. When things were going smoothly the round trip took about 24 minutes.

Ironworker Earl Calderon in the west anchorage. A good view of the way the wire loops around the strand shoes, November 2005.

The spinning wheel makes its way towards the west tower. Four bottom wires are dead, not moving, and fall into the cable. Four top wires are live, paying out quickly from a drum in the east anchorage, and will fall into the live wire sheaves on the cable former, March 2006.

The cable spinning was done in units called "strands." Spinning was done in two shifts working mornings and afternoons. A night shift then arrived to adjust the completed strands to their proper elevation. Adjusting was done at night so as not to slow down the spinning crews and because the cooler temperature made it easier to compensate for the expansion and contraction that occurs as the sun moves across the sky during daylight hours.

Night shift crew adjusting a finished strand near the west tower, November 2005.

A total of 19 strands, each containing 464 wires for a total of 8,816 wires, make up each of the main cables. As each strand was completed a crew worked the length of the bridge binding the strand with wire. Once all the strands were in place a large compacting machine was used to compress the octagonal shaped cable into a round shape measuring 20-1/2 inches in diameter and weighing in at six million pounds.

When spinning was completed on the south cable four 28,000-pound cable compactors were installed on the main cable. When the compactor was activated the six hydraulic rams, mounted in a circle and each capable of exerting 10,000 pounds of pressure, squeezed the cable into its round shape. Every two feet or so a one-inch stainless steel strap was wrapped around the cable to maintain its shape until the permanent cable bands could be installed.

Cable compactor working its way down the west side-span, Feburary 2006.

Ironworker "Cowboy" Butler attaches temporary cable strap on the west side-span, Feburary 2006.

Once the cable was compacted the installation could begin of the green cable bands that in turn support the suspender cables. Here Willie Perkins and Gerry Kerr mount one set of cable bands onto the main cable. Each cable had 132 such cable bands, May 2006.

By March 2006 the south cable is compacted and the cable bands are installed. Cable spinning continues on the north cable.

In March of 2006 crews began installing the first of 132 pairs of 1-5/8-inch wire rope suspender cables. Two suspender cables are draped over each cable band so four ends hang down to connect to the road deck. The suspender cables were delivered to the site on large wooden spools. The spools were then hoisted to the tower tops where the suspender cables were unwound from the spools and hauled to their position on the main cable. Like most everything else on the bridge, the suspenders were attached to winch lines and trundled out to the span using a work cart on the overhead tram lines.

Spooled suspender cable on the west tower.

Once the cable bands and suspenders were in place the main cable was coated with zinc paste and wrapped with a 1/8-inch galvanized wire to protect it from the elements. A canopy was used to shelter the work from the weather.

Detail shot of the cable wrapping machine and the wire. Over 900 miles of wire was used to wrap the main cables. When the wrapping was completed three coats of paint were applied to the cables and suspenders.

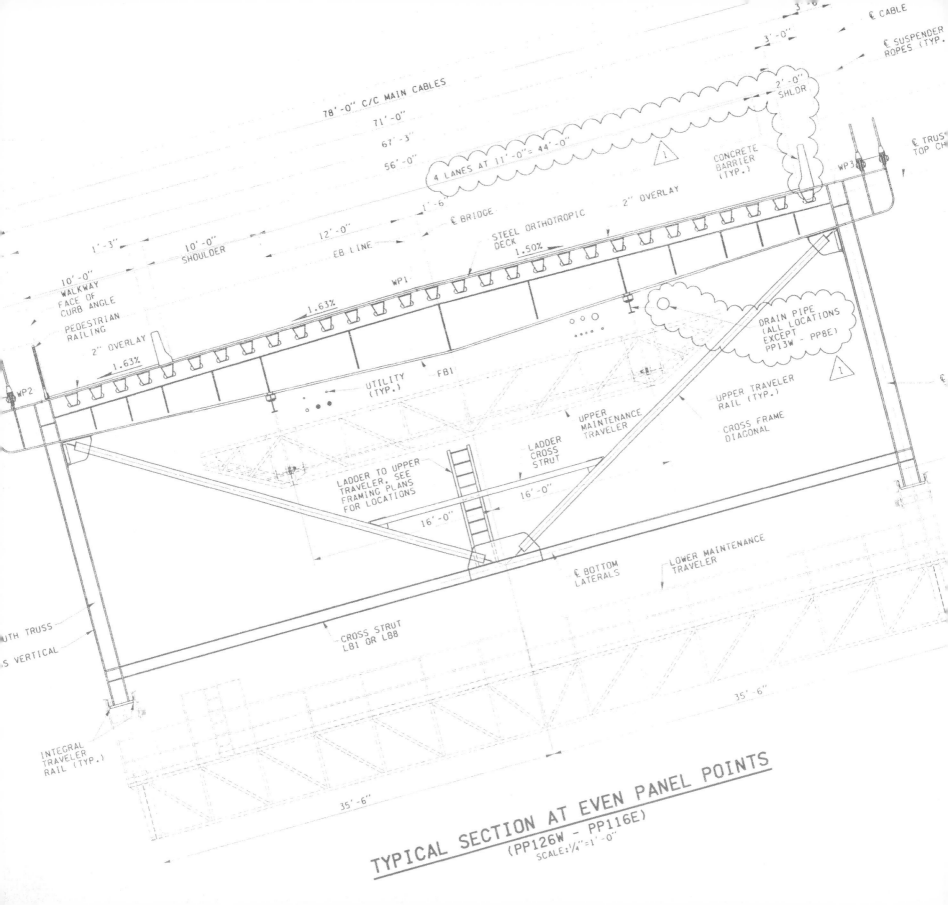

₵ CABLE

₵ SUSPENDER
ROPES (TYP.)

78'-0" C/C MAIN CABLES

71'-0"

67'-3"

56'-0"

2'-0"
SHLDR

4 LANES AT 11'-0"= 44'-0"

CONCRETE
BARRIER
(TYP.)

₵ TRUSS
TOP CH

WP3

1

1'-6"

₵ BRIDGE

2" OVERLAY

STEEL ORTHOTROPIC
DECK

1.50%

1'-3"

10'-0"
SHOULDER

12'-0"

EB LINE

WP1

10'-0"
WALKWAY
FACE OF
CURB ANGLE

1.63%

DRAIN PIPE
(ALL LOCATIONS
EXCEPT
PP13W - PP8E)

1

PEDESTRIAN
RAILING

2" OVERLAY

UTILITY
(TYP.)

FB1

UPPER TRAVELER
RAIL (TYP.)

CROSS FRAME
DIAGONAL

1.63%

WP2

UPPER
MAINTENANCE
TRAVELER

LADDER TO UPPER
TRAVELER, SEE
FRAMING PLANS
FOR LOCATIONS

LADDER
CROSS
STRUT

16'-0"

16'-0"

₵ BOTTOM
LATERALS

LOWER MAINTENANCE
TRAVELER

UTH TRUSS

S VERTICAL

CROSS STRUT
LB1 OR LB8

35'-6"

INTEGRAL
TRAVELER
RAIL (TYP.)

35'-6"

TYPICAL SECTION AT EVEN PANEL POINTS
(PP126W - PP116E)
SCALE:¼"=1'-0"

CHAPTER SIX

ROAD DECKS

Aside from the footbridge and the cable spinning one of the most visually arresting aspects of the new Tacoma Narrows Bridge was the arrival of the transport ship Swan under the bridge carrying the first 16 of 46 total deck sections that would make up the road deck.

Dockwise ship *Swan* sits under the new bridge ready to begin unloading the first of 46 road deck panels.

The mile-long roadway is made up of 46 deck sections. Built in South Korea, the deck sections arrived at the Narrows in three separate deliveries. The first 16 decks arrived in Puget Sound in June 2006. Averaging 120 x 78 ft (the shortest is 78 ft and the longest is 155 ft) and weighing in at roughly 450 tons, the deck sections were lifted into place by the eight blue gantry cranes that straddled the main cables. Two gantry cranes were stationed on each side span and four at main span between the towers.

Two gantry cranes in place near the Gig Harbor shore.

On the side spans, the gantries' lifting mechanisms were powered by large winches located on the pier tops at the foot of each tower. In the mid-span area, the gantry cranes lifted the sections using strand jacks mounted on the gantries themselves. Each section was attached to vertical suspender cables, and then connected to the main cables.

Detail of main span gantry crane with strand jack and cables spooled in green housing, July 2006.

Detail of gantry care on west side-span. The gantries had clamps to grip the main cables and a hydraulic mechanism that allowed them to walk up and down the cables and step over obstacles like the cable bands, December 2006.

The first deck was placed at the center span, the middle of the main cable, on the evening of August 7, 2006. All of the deck sections were lifted off of the ship near the Gig Harbor shore and loaded onto barges. Once on the barge the deck would be moved into place at slack tide when the water was calm. Tugboats and four independent motors attached to the barge kept the section in place while a picking beam was attached and the gantries could lift the deck.

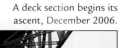
Surveyors monitor the progress of one deck section as it is lifted into place.

A deck section begins its ascent, December 2006.

The 46 deck sections were raised in a planned sequence designed to maintain equal stresses on the bridge's two towers and distribute the weight across the main cables evenly.

The weight of all 46 sections pulled the main cables down about 25 feet below their working profile from when the footbridge was constructed and the main cables were spun. This deflection process placed the new bridge deck at the same level as the 1950 bridge deck, approximately 190 ft above the water. As the deck sections were put in place crews bolted the trusses and later welded the deck plates.

A view looking east from under the deck sections.

View from west tower of installed deck sections and transport ship in the channel below, December 2006.

When the deck was completed it formed a continuous 5,400 ft roadway with expansion joints at the east and west anchorages only. Once the deck sections were bolted and welded the surface was topped with two inches of asphalt.

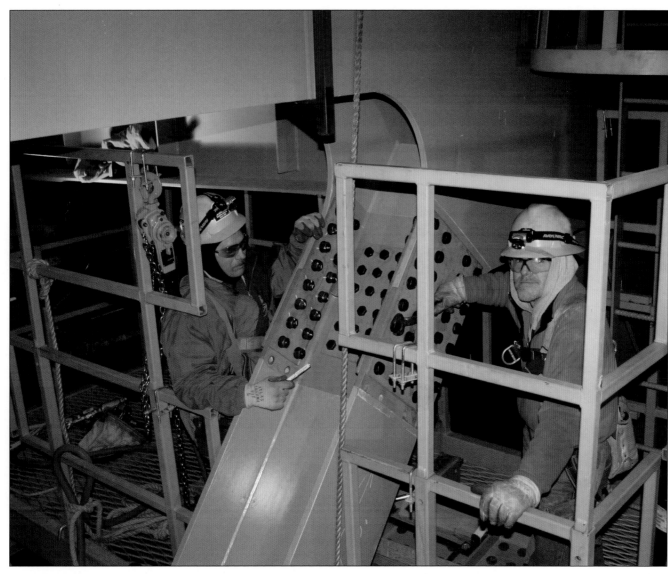

Ironworkers bolting the trusses of the deck sections, December 2006.

Looking towards the west tower from center span. The catwalk is festooned with Christmas lights. The tent across the road deck allowed workers to weld the deck seams shielded from the elements.

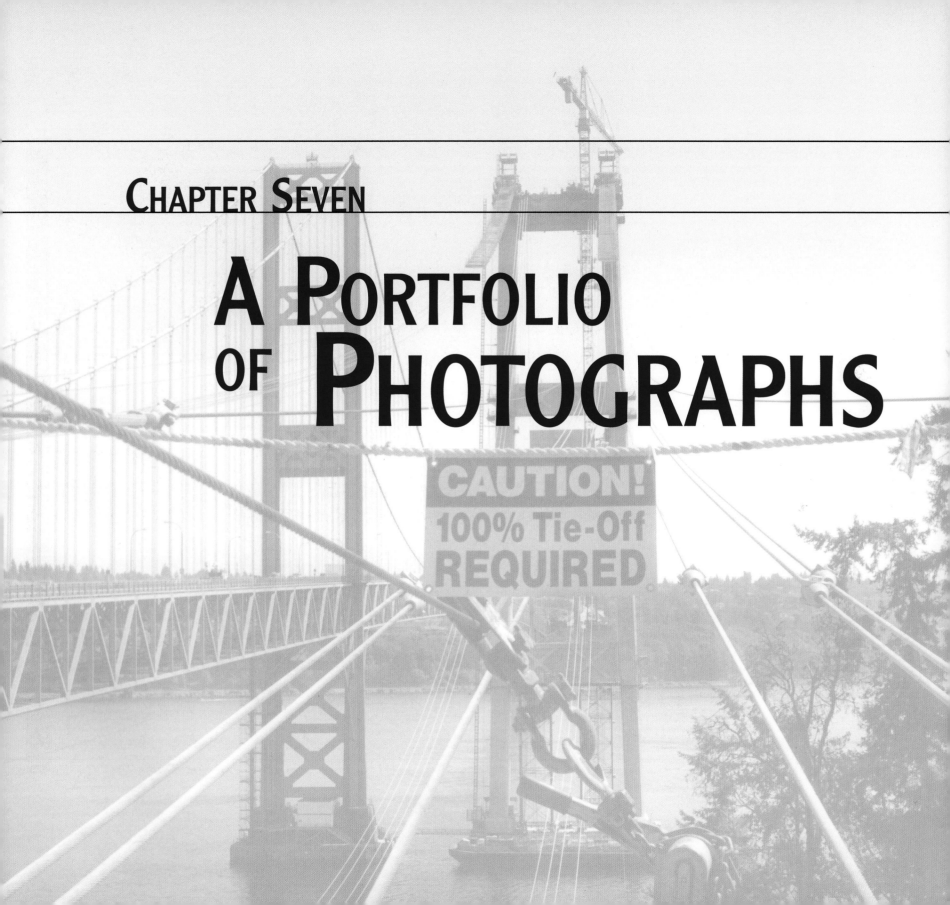

CHAPTER SEVEN

A PORTFOLIO OF PHOTOGRAPHS

When I started this project I shot mainly black and white film. About half way through I started using a digital camera for color work. While I still favor black and white film I am increasingly impressed with the quality of the digital images and the camera's ease of use. Over the past several years many hundreds of people have worked to make this bridge a reality. Engineers and architects dreamed it up and designed it and the building trades people built it. The following pages gather photos of people, work, general site shots, and abstract images to give a visual overview of this remarkable project.

Spools of hold-back cables near the Gig Harbor anchorage, April 2005.

Clearing storm at dawn, looking east from the west tower, May 2005.

Looking up at the west tower with bird cages almost to final height, May 2005.

Ironworker, Dan Weed, on the rebar cap to the west tower's north leg. Bridge workers called the cap the "wedding cake," May 2005.

Laborer, Shelly McConville and Erin Hunter, TNC Public Affairs Manager, atop the east tower bird cage, May 2005.

Rising to over 14,000 ft Mount Rainier presents an impressive backdrop for the east tower in the afternoon sunlight, May 2005.

Ironworkers on the footbridge early in the construction of the west side-span, August 2005.

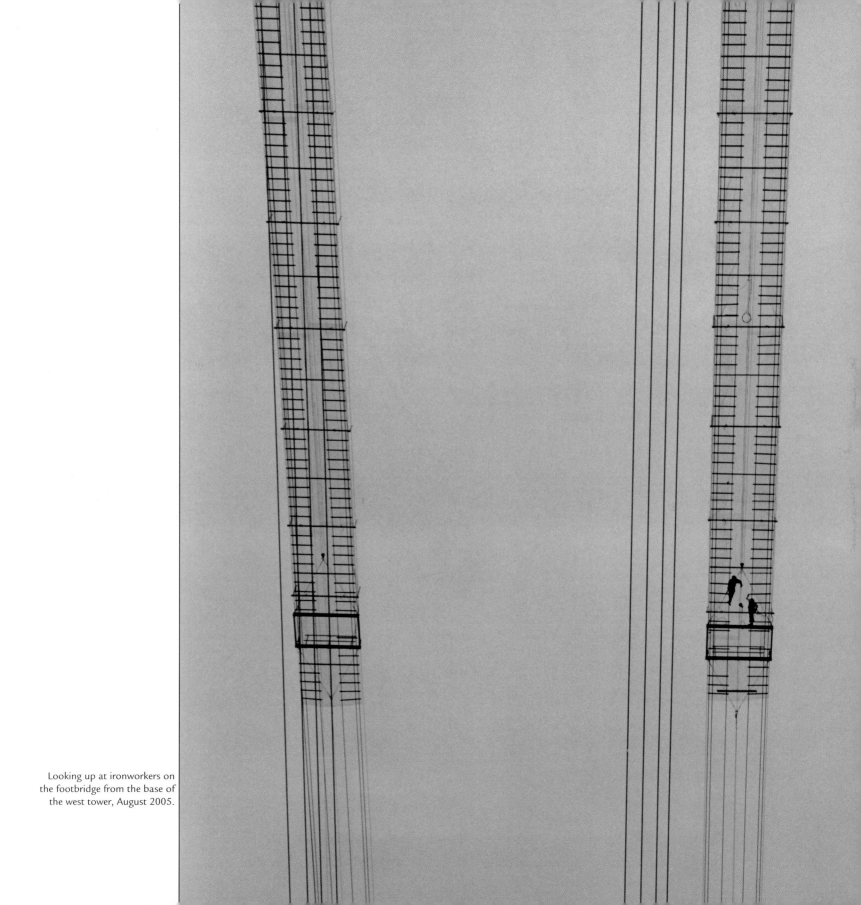

Looking up at ironworkers on the footbridge from the base of the west tower, August 2005.

Crane operator, Richard Sokolik, inspects the crane's jib, August 2005.

Skyler Willard on the west tower of old bridge. Footbridge is in early stages of
construction in the background.

Closing the footbridge at main span, September 2005.

From left to right: Marty Lenoir,
Tony Benson, and Brady Cooper,
at midspan, September 2005.

Frank Rosenberg and "Cowboy" Butler on the south footbridge as it reaches the middle of the main span, September 2005.

The new bridge rises out of the fog next to its older sibling, September 2005.

Superintendent Tracy Martin,
August 2005.

Operator and winch
atop the west tower,
August 2005.

83

From left to right: Dave Brown, Justin Wiedrich, and Steve Peterson at center span, October 2005.

Workers securing one of the nine cross bridges that allowed access between the two catwalks. The cross bridges were hoisted into place with lights and other electrical gear used to light the catwalks for night work, September 2005.

Miles Coble on the east tower work platform unrolling mesh for the footbridge, September 2005.

T.J. Johnson atop the east tower.

Dave Holcomb and Richard Smith in the east anchorage as work proceeds on the first strand, September 2005.

Scott Minton at the east splay saddle during spinning of first strand. Four live wires are in the live-wire sheaves to the right, October 2005.

Looking up at the east tower early in cable spinning operations, November 2005.

View looking west from the
tower crane, February 2006.

Jason "the Great" using a shepherd's hook to guide the wires into their proper position, March 2006.

Spinning wheel and eight parts of wire coming into the Gig Harbor Anchorage, October 2005.

Moving the compactor up the main cable after completion of the cable compacting.
Here the crew is moving the compactor past one of the tram frames.

Another view of workers moving the compactor under a tram frame. Four electric hoists mounted on the tram lines were used to support the compactor on its trip up and down the cable. As the crew came to a tram frame they had to move each hoist around the frame.

Brady Cooper preparing to remove the compactor from the main cable.

Ironworkers at the west tower top rigging the compactor for removal after completion of the cable compacting.

Todd Murray aka "Big Daddy"
on the west tower, August 2005.

Ironworker general foreman Jeff Glockner at mid span.

Ironworkers drag a winch line to center span to remove the last of the tram-frames used during cable spinning, May 2006.

Ironworker Steve Rininger atop the west tower.

Gerry Kerr, "Kerr Dog,"
on the west tower.

Footbridge and tower at night, October 2005.

Ironworker Casey Goings near the gantry crane on the west side-span. Casey, a talented artist, drew numerous caricatures of his fellow bridge workers.

Scott McMullen rigging one of the pendants used to trapeze some of the deck sections into place. Scott is an evacuee of Hurricane Katrina. He left New Orleans with two days notice and came to Tacoma where his wife's company had an office. He reported to the Seattle hall and was put to work on the bridge. As for his future plans he says, "I've got nothing to go home to. It was all wiped out. It's like I never existed there," June 2006.

Ironworker on the west tower work platform signals the winch operator to begin trundling a cable band down the footbridge into its position on the main cable.

Operator Tina Keller at the wire spooling station near the Tacoma Anchorage, November 2005.

Parriss Stull moves one of the pendant cables into place on the east side-span, June 2006.

Engineer Roy Wilson on the east tower in May of 2005. An experienced bridge engineer, Roy has worked on suspension bridges around the world. Before coming to Tacoma, Roy worked in California on the Alfred Zampa Memorial Bridge.

Ironworkers work from both the footbridge and the cross bridge to slack-out the last of the holdback cables, May 2006.

110

A weld bay was set up in the west anchorage to train and test the welders who would weld the critical road deck seams, November 2006.

Operator Elton Plank operating the pier-top winch that controlled the lifting of the deck sections on the Tacoma side-span, December 2006.

Decks in flight.

TNC Project Manager Manuel Rondón on the stair tower overlooking the new bridge, May 2007.

Bridge at twilight on Super Bowl Sunday 2006. The catwalk is decorated in the Seattle Seahawks team colors in honor of the team's appearance in the Super Bowl.

Thick fog enshrouds the bridge almost to the tower tops. The fog is a visible symbol that bridge builders don't just contend with the laws of physics but with the elements as well, February 2006.

Spool of wire near the east anchorage,
October 2005.

Night view of the rigging used
to tension the hold-back
cables, November 2005.

Looking up at the main cable as it splays out to the various strand shoes in the Tacoma anchorage, May 2006.

Clayton Glavin atop the Tacoma tower waiting for the wheel to return, October 2005.

Parriss Stull tightening a cable band on the Tacoma side span, May 2006.

While the new bridge was being constructed the old bridge was getting structural upgrades. Here the retrofit crew poses for a photo on the walkway under the existing bridge. From left to right are: Ron Howell Sr., Peter Cannata, Steve Iwakiri, Patrick Grady, and Michael Stengle, October 2006.

Detail of rigging and spooled suspender cable on the west tower top, July 2006.

Footbridge at night looking east from the Gig Harbor shore, October 2005.

Superintendent "Mad Dog" Dan Maddux on the Gig Harbor anchorage, October 2005.

Brady Cooper marks the wires with yellow and black marking pens at the west anchorage. The wires were color-coded to help organize their placement in the cable strand, November 2005.

Night shift crew coming off the catwalk, November 2005.

Ironworker Ron Carrier at the start his shift, November 2005

Night shift foreman Jim
Kostelecky ignores the
posted prohibitions.

T.J. Paul and Sean McCormick near the west tower prepare to adjust one of the cable strands, November 2005.

In March of 2006 retired ironworkers were given a tour of the bridge. A platform constructed next to the Tacoma anchorage gave visitors to the bridge a good view of the cable spinning process from safe a vantage point. Some of the men on this tour worked on one, or both, of the old bridges.

Retired ironworkers at the bridge, from left to right are: Rudy Lutz, Doug Smith, Willie Hyduke, Gene Myers, Keith McClelland, Earl Bachman, Reg Carson, Roger Kiplinger and Earl White.

Crew working on the main span adjusting first strand of north main cable, October 2005.

Ironworker installs cable band on the main span south cable while spinning work continues on the north cable, March 2006.

Removing the sheave battery from the east tower top, March 2006.

Graffiti is as old as construction. Graffiti inscribed by work gangs has been found in the Pyramids. A common motif on the bridge was this skull and crossed spud wrenches—a tool specific to the ironworker's trade.

Tower crane operator Richard Sokolik (left) and operator Thom Schell on the east tower at the start of their shift, February 2006.

Electrician wires the cable compactor in preparation to begin compacting the Tacoma side-span. Before compacting the cable had a loose hexagonal shape. This shape is achieved by spinning strands in groups of 7, 19, 37, 61, ect. The new bridge has 19 strands of 464 individual wires in each cable. While, for example, the Alfred Zampa Memorial Bridge in California has 37 strands each containing 232 wires in each cable.

Tugboat moves tower crane pieces into the west tower, February 2006.

Suspender cable on spool at the foot of the east tower, May 2006.

Bridge crews install one of the last suspender cables near the east tower, August 2006.

Looking east towards the Tacoma shore. The suspender cables are installed on the south cable, May 2006.

Abstract detail of suspender cables, May 2006.

The *Swan* viewed from the top of the west tower, July 2006.

Dockwise ship *Swan* stationed under the bridge with first load of deck sections, July 2006.

The 19th century meets the 21st century, in August of 2006, as a ferryboat moves past the first deck section as it is lifted into place.

First deck section in place at midspan, August 2006.

Deck section begins its ascent as the barge and tugboat move away in the distance. The barge has four blue motors used to stabilize it under the bridge as deck section is being attached to the lifting beams lifting beams, December 2006.

Deck section one in place at the Gig Harbor anchorage, September 2006.

A view of the bridge deck leading to the Gig Harbor anchorage from the stair tower
leading down to the pier of the old bridge, November 2006.

Deck sections were brought to the shore via a process called trapezing. This process involved transferring the load from the gantry cranes to suspended pendants then repositioning the gantry cranes, repicking the load, and transferring it to another set of pendants positioned further ahead. This shot from October 2006 shows the pendants taking the load from the gantry lines which are at a near 45 degree angle.

Mount Rainier and the bridge with gantry cranes midway up the side-span, December 2006.

Ironworkers dismantling the north catwalk, May 2007

Painters aloft, May 2007.

Looking east from the Gig Harbor shore the bridge is near completion, May 2007.

Looking west from the east tower at the nearly completed bridge. The north footbridge is being disassembled, May 2007.

ABOUT THE AUTHOR

John V. Robinson, a member of Iron Workers Local Union 378, is a photographer and writer educated at U.C. Berkeley and San Francisco State University. John is the author of several books including *Spanning the Strait: Building the Alfred Zampa Memorial Bridge* (2004).

John lives in California with his wife Lisa and their three kids: Kyle, Kathy, and Ian. In 2006 he was awarded a Guggenheim Fellowship to pursue his documentation of bridge builders. John wishes to thank the Guggenheim Foundation for supporting his work.

Spanning the Strait
by John V. Robinson

116 pages 120 photographs & Illustrations
Paperback $24.95
ISBN 0-9744124-0-6

Spanning the Strait: Building the Alfred Zampa Memorial Bridge combines an oral history of legendary iron worker Al Zampa with a photo-documentry of the new suspension bridge named in his honor.

To Make the Run
by Joe E. Gladstone

284 pages 18 Illustrations
Paperback $24.95
ISBN 0-9744124-2-2

In *To make the Run* retired merchant sailor Joe Gladstone combines personal anecdotes, traditional sea lore, and political philosophy, to create a compelling and lively sketch of a sailor's life during and after the Second World War.

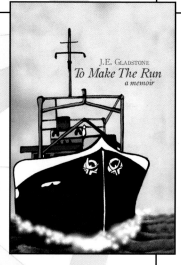

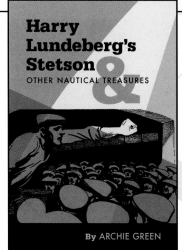

Harry Lundeberg's Stetson & Other Nautical Treasures
by Archie Green

155 Pages 24 Illustrations
Paper back $24.95
ISBN 0-9744124-3-0

In *Harry Lundeberg's Stetson & Other Nautical Treasures* Archie Green explores the history and nautical lore of the Sailors' Union of the Pacific.

Carquinez Press
P.O. Box 571
Crockett CA 94525
510 334 4269